Scottie Crews Media, LLC Presents:

Ellen E. Goes Fishing

Cameron

Always Believe!

G.S. Crews

Written by **G.S.CREWS**
Illustrated by **Chloe Vicos**

Printed in the United States of America.
For ages 6 and up | Grades 2 and up

Ellen E. Goes Fishing
ISBN: 978-0-9795236-6-3
Written by: G.S.CREWS
Published by: Scottie Crews Media, LLC
Book layout by: Scottie Crews Media, LLC
Illustrated by: Chloe Vicos
The illustrations were created by Atlanta-based artist, Chloe Vicos, who is an established illustrator.
You can vist her online at www.instagram/craisin_bran.
Edited by: Windy Goodloe
The book was edited by Windy Goodloe. You can contact her at windy.goodloe@gmail.com.

SCOTTIE CREWS
MEDIA

www.scottiecrewsmedia.com

Dedication

This book is dedicated to Ellen Gaston Crews. She was her five sons' first teacher. Your lessons live on and so does your memory.

Hi! My name is Ellen E. The "E" stands for Engineering, but everyone calls me Ellen E. for short. I am eight years old. In dog years, that is fifty-six years old. Jumpin' Jehosaphat!

My dad and I went to the Big Store. We were shopping for some fishing gear before he took my friends and me fishing at the lake!

Who are my friends? Well, they are none other than Sally Science and Maria Math!

When we stopped on the sporting goods aisle, my dad held up a rod. He asked, "Ellen E, do you know how this fishing rod and reel were made?"

"I'm sure it was made in a factory," I said.

"You see, once upon a time, a girl in a little town wanted to catch some fish, but she did not have a net."

"I see where you are going with this."

"One day, she was messing around in her father's basement when she tripped over something. When she looked over, she saw a rod and had a fantastic idea."

"Jumpin' Jehosaphat! I don't know where you're going with this."

"That night, the little girl pulled out her notebook and sketched a picture of a rod and reel. She'd made a prototype."

I pulled out my notebook and held it high in the sky.

"I know what a prototype is! A prototype is a model that isn't ready to be seen by the world yet! Just like my top-secret, super-duper mind-reader device! Once I finish it, it will be able to tell what you are thinking!"

"That's right," said my dad. "Now open that tackle box and look inside."

I opened the tackle box and saw many colorful lures inside.

"Heavens to Murgatroyd," I shouted.

"Each of these lures have a specific weight and length. Each of these designs started as someone's imagination," said my dad. "Your imagination is the fuel for your ideas. Every engineer needs an imagination, so they can create new prototypes."

"Jumpin' Jehosaphat!"

After my dad bought some fishing rods and tackle boxes, we left the store and went to pick up my friends. I was excited about going fishing at the lake!

Sally Science was waiting at Maria Math's house. When we pulled into the driveway, they were waiting on the front porch. Like always, Sally Science was wearing her white lab coat, and she had a pink bow in her curly hair.

Sally Science always carried an earth science journal that she used to help her explain scientific things.

Maria Math liked to wear her book bag with her GT T-shirt and her blue jeans. She carried a tablet in her book bag. My mom always said she was a "numbers" woman.

My dad stepped out of his SUV and greeted my friends, "Hi, kids!"

"Hola, Mr. Scott," yelled Maria Math.

"Hi, Mr. Scott," yelled Sally Science.

"Are you ready to go fishing?" asked my dad.

"Yes, Mr. Scott," Maria Math and Sally Science yelled at the same time.

"Then, hop in!"

My dad opened the back door for my friends. I got into the front seat. We put on our seatbelts for safety before my dad pulled off.

"Dad and I just came from the Big Store," I said to my friends. "We got something for you all."

"What did you buy?" asked Sally Science.

"We bought fishing rods and tackle boxes. There are colorful lures on the inside!"

"Hip! Hip! Hooray!" cheered Sally.

"Excelente!" cheered Maria Math.

My dad slowly maneuvered the SUV out of Maria's driveway. As he drove us to the lake, Maria Math pulled out her tablet.

"Ellen E., does your dad's car have Wi-Fi?"

"No, but I can create a hot spot with my cell phone," I replied.

"Cool. I want to log into my school's SharePoint and teach you how to multiply by tens."

"What is SharePoint?" I asked.

"SharePoint is where my teacher shares all our homework assignments and vocabulary words," Maria Math said. "All we need is a Wi-Fi connection to help us log in."

"Jumpin' Jehosaphat! Give me zero to seven seconds!"

I pulled out my cell phone and created a hot spot. When Maria Math's tablet recognized the network, I entered my password: JUMP.

The tablet connected. Then, Maria Math put in her username and password for SharePoint.

Voila! We were in!

Maria Math began to teach us.

"¡Estudiantes!"

"That's Spanish for 'students,'" I whispered to Sally Science.

"¡Escuchar a la maestra!"

"The official translation is 'listen to the teacher,'" Sally Science whispered back.

"When you multiply by ten, add a zero to the other number. What is ten times two?" Maria Math tapped the surface of her tablet, and the math problem appeared.

"Ten times two equals twenty! All you do is add a zero to the two," I said as I gave Sally Science a high five.

"Well, fifty times ten equals five hundred!" shouted Sally Science.

"Si!" Maria Math said.

"We are so smart! So smart! So smart!" chanted Sally Science as we all celebrated.

The SUV stopped. We had arrived at Lake Shamrock.

After we leaped out the SUV, my dad gave us our fishing gear. We ran to the edge of the lake where we could see the fish leaping out of the water as if they were happy to see us.

"Jumpin' Jehosaphat! Just looking at all this water makes me thirsty!" I said.

"About 70 percent of the Earth is covered by water," Sally Science said as she reached into her lab coat and extracted her earth science journal.

"Sally, can you tell us the secrets of water?" I asked.

"I absolutely, positively can," Sally Science said proudly. "Do you know why every living thing needs water?"

"Wait! Let me take notes so we can share them later!"

I sat at the picnic table with Maria Math. I pulled out my notebook and was ready to take notes.

Sally Science began to speak.

"The cells in our bodies need water to survive because water transports nutrients through our bodies."

"Oh! So that's why we need to drink eight glasses of water every day," Maria Math said as I got up and did a cartwheel.

"That is absolutely, positively correct! Now, Ellen E., quit doing cartwheels," requested Sally Science. "There are more facts that I need to share."

After I did one last cartwheel, I went back to my notebook.

"Water has three states: liquid, solid, and gas," Sally Science educated. "The states of solid and gas occur at two specific temperatures."

"Can you tell me the numbers at which these states occur?" asked Maria Math.

"I absolutely, positively can. Water freezes solid at 32 degrees Fahrenheit, and it boils at 212 degrees Fahrenheit where it becomes a gas!" Sally Science said.

Splash! Splash! Splash!

We all turned around to see my dad reeling in a large fish! It splashed water here and there!

"I know you are not going to let me catch more fish than you," my dad said as he looked at my friends and me.

"No, sir!" we all said.

We picked up our fishing rods and cast our lures into the lake. Using technology to learn about engineering, science, and math is fun. Jumpin' Jehosaphat! I can't wait to learn more!

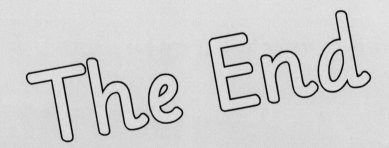

Ellen's Catch of The Day

What is a prototype? (Page 9)

What is SharePoint? (Page 13)

Translate "¡Escuchar a la maestra!" from Spanish to English. (Page 17)

About what percentage of the Earth is covered with water? (Page 19)

What are the three states of water? (Page 21)

At what temperature does water freeze? (Page 23)

At what temperature does water boil? (Page 23)

ABOUT THE AUTHOR

G.S.CREWS is a highly skilled author who has published six novels. He uses his ability to immerse his readers into the story line, causing them to become attached to the lives of the characters.

His purpose in life is to reduce the staggering number of illiterate Americans that read below a 5th grade level.

G.S.CREWS is a graduate of Clayton State University and he resides in suburban Atlanta.

You can find out more about him by visiting his website at www.scottiecrewsmedia.com.